Gravitation

& Our World

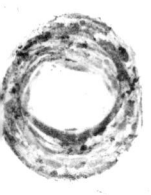

Franz Sundfeld

For

my daughter

Silke

with my gratitude for her
valuable help editing this book.

I postulate

1st. Edition

Preface:

Already ancient Greek philosophers - one was Thales from Milet and his followers, - were of the opinion that the whole world has emerged und has been constructed out of a "Unique Primal Substance".

This idea of a "Unique Element" of the nature still exists today and it is the reason of a zealous endeavor to find that "Basic Germ". Once this Primeval Substance is known, it is believed that we can be able to unravel the old dream of human kind: to understand how the Universe was formed. Therefore, it is currently of scientific interest and utmost importance, to lift the curtain that prevents us to take a look at this strange and mysterious "Something" and to recognize finally the unique truth.

"At the very beginning next to matter and radiation also space and time was created in the Big Bang". - Says G. Börner [1], *and continues: "If we follow the physical theories into their deepest consequences, then we see that they need to be supplemented, if space and time have been formed in the big bang and vanish into the black hole, then our well-ordered world of time and space cannot be everything ... but our theories show us that we need ideas that go beyond time and space to gain a full explanation."*

This essay is an attempt to move into this unexplored territory.

Introduction

From the formula of Einstein we know that energy and matter are inter-convertible and assuming that there are in the Universe two Realms: the Energy one and the Matter Realm. The former consists of a singularity - most concentrated Primal Energy or gravity and the second is the spatially far extended medium of all the matter - consisting of the us so well-known world, with the galaxies, the sun, the stars, planets and moons, elements, radiation, the space and time.

The Universe is "alive" and there is a constant transition from Primeval Energy into Matter and vice versa. In order that this interaction can be maintained, each realm has its assigned location.   The Primeval Energy forms the spherical core of the Universe, consisting of a singularity, which with its enormous gravity binds the outer realm or the outer world and constantly keeps it under control. Both areas hold themselves in a slightly oscillating balance; with the constant conversion from one state to the other.

The inner core must be protected from the "outer world"; both realms must not be in direct contact, this provides an Interlayer. This must, above all, prevent that the temperature of the matter in the outer world has direct contact with the Energy Core; this would otherwise begin to expand, since only at zero degrees K it is completely stable. As we know, the intergalactic space of matter has a temperature of 2.73 degrees K, which is called the Background Radiation. The Interlayer is transparent against the strong effect of the gravitation of the core on the matter in the outer space and above all it allows the constant exchange of the Primeval Energy, after the change of the phase transitions, but keeps the temperature away from the core.

It should be mentioned where exactly the change takes place:

In the Black Hole, "Matter" is converted into Primeval Energy by the compression process and this is absorbed by the singularity of the core.

In Quasars new "Matter" is created from a "Part" of the Primeval Energy, the singularity of the core.
With that the circuit is closed.
Steven Weinberg defines already both objects: Black Holes and Quasars in this way in his book "The first three Minutes".

THE STRUCTURE OF
THE GRAVITY-UNIVERSE

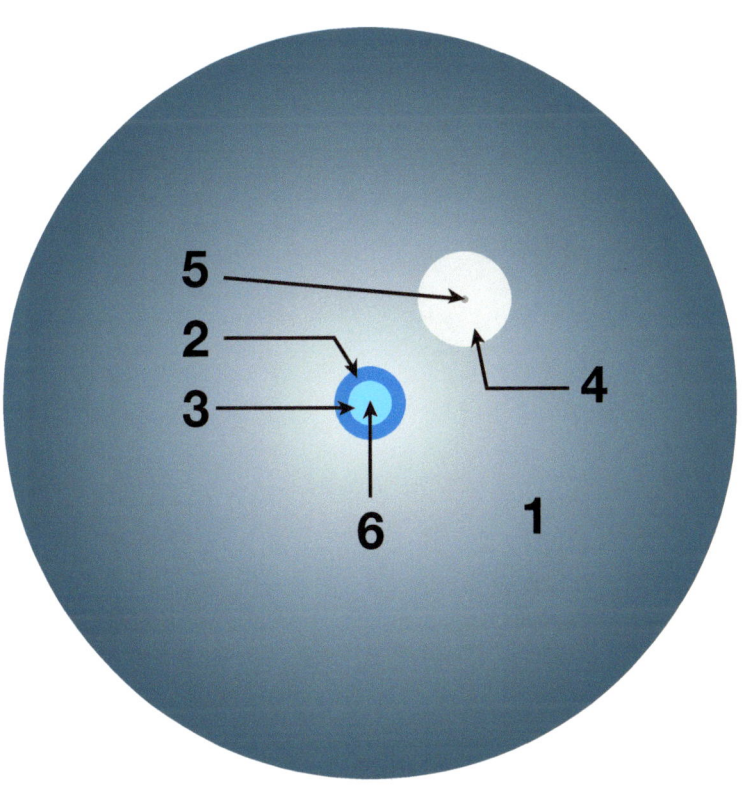

1 **Matter Realm**
2 **Interlayer**
3 **Primeval Energy Core**
4 **Visible Part of the Universe**
5 **Centre of the Visible Part of the Universe**
6 **Centre of the Gravity-Universe**

Drawing 1

ENERGY CIRCULATION

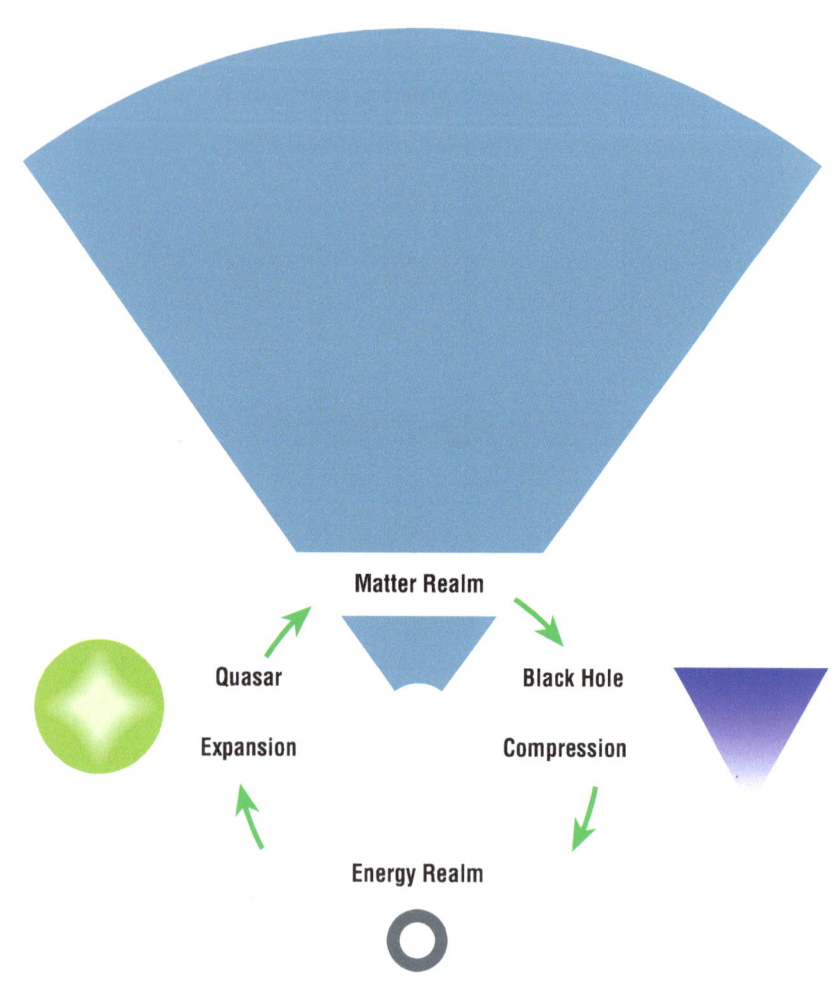

Matter Realm

Quasar   Black Hole

Expansion   Compression

Energy Realm

Drawing 2

Explanation Drawing 1 & 2

If we now come to the actual topic of the origin of the "Gravity-Universe", an important step has to be cleared regarding to the conversion of Matter into Prime Energy. This is, therefore, of so great importance, for just as on one hand, the vanishing of Matter is going on, the formation of the same must also take place however in the reverse order.

In 1995, the physicists Eric Cornell and Carl Wiemann first produced a Bose-Einstein-Condensate from rubidium gas in the laboratory. They have been able to observe the behavior of the atoms at the lowest temperatures, just before the absolute zero point, and they have discovered that the atoms remain motionless and at a few billion of degrees above 0°K, the Matter dissolves and becomes a confused mass, all atoms are "smeared" forming a so-called "Super-Atom". At exactly 0°K, the last act of the transformation has to take place and the Super-Atom is transformed into Primeval Energy (a singularity).

This last step has already been scientifically mentioned, which is to be shown here: On the Internet, the term "NEUTRONIUM" 2) - the term used first by Andreas von Andropoff in 1926 - refers, among other examples, to the conversion of "neutronium" in the neutron stars at highest pressure:

"The high density of neutronium also generates a high and very strong gravitational field that it may cause the neutronium core to collapse in on itself, forming a gravitational singularity, (colloquially known as a fl. hole)".

This consequence is the following step in the transformation of Matter, on its transition into a singularity = Primeval Energy. In any case, we see here the exact process of the "vanishing" of Matter.

Subsequently, the transition product is called Neutronium-Condensate. Taking into account the just mentioned specifications, the actual topic of this work can now be carried on.

The Universe, from its origin to its finished function

Mankind must by its very nature resign itself to certain obstacles which do not allow taking a look behind them. So, nobody can say what was before his birth, where he comes from, or where he goes after his death. Nor can anyone tell where the singularity came from, of which the whole Universe emerged. Even the scientifically accepted standard model of the Big Bang can`t give us an answer to the first two basic questions:

1 - Who created, or from where came the original singularity?
2 - What caused this just in the precise moment, to initiate the phase transition of Energy into Matter?

Regarding to the first question we stand before the mentioned certain barrier, which hardly ever allows us to look behind it. As to the second question, an acceptable solution might be found. Later on we'll refer to this.

Again: the "Gravity-Universe" consists of two separate realms: the matter one and the energy core, which is a singularity. The spherical nucleus forms the center of the Universe, around which the huge "outside world", the matter sphere extends. The Universe has emerged from an initial singularity. This **singularity** consisted of pure **Primeval Energy** or **Gravitation** In this transformation process no other substance, (energy, matter or some other particle) was implicated – it was just and **only** the singularity, that expanded into pure matter. Therefore we can assure:

**SINGULARITY CONSISTS OF PRIMEVAL ENERGY =
HIGHEST-COMPRESSED GRAVITATION.**

**MATTER CONSISTS OF EXPANDED PRIMEVAL
ENERGY = EXPANDED GRAVITATION.**

**SO, GRAVITATION IS BOTH THE CORE-SINGULARITY
= AS WELL AS MATTER.**

CONSECUENTLY GRAVITATION HAS

TWO PHYSICAL STATES:

On one hand, it is a SINGULARITY = pure ENERGY

And on the second hand, it is MATTER.

So, our entire world system consists only of one basic

substance:

GRAVITATION

This is affirmed by itself, both as matter and as singularity, by its constant kind of action, its ATTRACTION, which can't be neutralized, since it is exactly the same basic force in both phenomena.

We have been familiar with the Big Bang Theory for decades. Its pure singularity or gravitation developed into our world-wide known matter. Here we already have the evidence that gravitation is the same in both cases: singularity and matter. **This again confirms the two physical states of gravitation**.

Now the development of this "Gravity – Universe" model is to be described in individual pictures.

The two basic forces

The whole process in the Universe is determined by two forces: a positive one represented by energies and matter that was once united in a singularity, and resulted in the whole, ever existing gravitation, with the "inward" directed force concentration. The second is an "outward-facing" equivalent negative "force" which is completely "empty" harboring no energy or matter in any form, and precisely for this reason embodies a counter-momentum to the former, and represents the greatest conceivable emptiness- the vacuum.

These two opposite features determine and control everything that is going on in the entire Universe and what should ever happen. The vacuum, however, is not infinitely large; it extends as far as the gravitational field of the nucleus still acts, and forms with it a closed unit. We ignore what's going on further outside. This is at any rate the closed sphere of our world, a completely centrally oriented and dominated entity. Both counterbalances are - and this is fundamental - at 0°K in a state of equilibrium: the gravitation of the singularity concentrating on the center and the "emptiness" of the surrounding vacuum acting in all outside directions. The singularity could fill the vacuum of the Universe with matter and radiation, but it doesn't, because of the strong, counteracting gravitational forces of the nucleus. The vacuum is "hungry" for the slumbering singularity however gravitation does not allow bringing it out. Thus the singularity has always hovered in space without any change in the state of equilibrium, darkness, icy coldness, silence, and extreme calm. The key factor here is the absolute absence of temperature.

Imagine if the vacuum could affect the core, matter would flow into it and fill it. However the whole matter of the singularity would not flow into the vacuum, but only part of it, until a new balance has been established. The expanded part of the singularity diminishes the void of the vacuum, but the most important thing is that the gravitation of the rest of the singularity balances now with that of the emitted matter, and the vacuum must be satisfied with it. There would therefore remain about half of the singularity as such, although the vacuum can still absorb much more matter in the vastness of its space. In this state of equilibrium, the great emptiness the "hunger" of the vacuum never ceases! This fact is fundamental; it is an important component in the world called the "Black Zero Energy".

We also have to mention the possibility in which the whole singularity would flow into the vacuum. In that case we would have not only a disturbed balance of forces, the gravitation would be no longer divided, one part in the core and the other part in the matter itself, but now the entire gravitation would dominate the matter. The consequence would be that here the gravitation would be so strong, that the matter collapses under its own weight and would very soon end up in black holes. This process would continue until a new state of equilibrium had occurred between the newly emerging singularity in the nucleus and the remaining matter in the space.

Now we come back to the two great and already mentioned unknown items. The first question, where the singularity came from, does not stand alone, the equally important uncertainty is regarding to the impulse which excited and activated the slumbering singularity, to be transformed into a whole Universe. Without this excitement the silence would remain, and the state of the frozen energy would be preserved forever. Actually, we would have to ask a higher authority to end this condition. But perhaps we are able to find an appropriate solution.

The Trigger

Let us assume that at the beginning there was not only one singularity but two of them, and that both moved with an enormous distance in space, until one came too close to the other and was absorbed by it. Basically this does not result in any significant reaction, because energies merge without any friction and thus do not develop heat or other effects. The result was only one single, greater singularity which continued to stay at 0°K in the vacuum. Nevertheless this caused an imbalance of "forces". The inward gravitational force of the new nucleus increased significantly, thus incrementing the force difference from the latter to the vacuum. This disturbing imbalance had to be eliminated in order to restore the former state of equilibrium. In the former two singularities this status quo existed, as long as their spheres of influence were separated: each one had a smaller energy concentration a smaller radius, and, above all, a very slow rotation.

How then was this newly - emerged disturbance to be eliminated?

1 - The first singularity separates from the second one by "rejecting" or releasing it. This is absolutely impossible since the two singularities are not only inseparable but their two gravitational energies would immediately merge again.

2 - After the moment both singularities came together a defrosting procedure started just beneath the surface of the core which initiated the formation of new matter. Now this process will be explained in detail.

First signs of Matter

Before the fusion of the two singularities, their orbits ran helically together, and after the collision the resulting larger formation received a heavy impulse which started an enormous rotation.

This spin and the combined gravitations of the new singularity very soon settled on a common center, so that it assumed the spherical form which is customary in the cosmos, naturally with a double-strong, centrally oriented gravitational force. However, the axis of rotation has no fixed orientation at the beginning, it fluctuates over some time.

And soon arouse in the new structure which has just emerged from the two singularities, a significant, but quite normal, side-effect, which however resulted in a fundamental change. The rotation of the core which has just initiated, has produced a new phenomenon, besides of the immense gravitation, a centrifugal force appeared. Although the centrifugal force is only a locally occurring phenomenon and can´t be compared with the powerful gravitation dominating the entire Universe, its effect is enough to bring about a controlled expansion process. Under the surface of the core, this force produces an "expansion" or "loosening" in this area of the singularity, which is the first step towards the beginning of a phase transition. In this peripheral area the loosening effect started an appearance of tiny bubbles that rise to the surface of the core and cover it with a "foam layer". This foam consists of the super-atom, the bulky, smeared mass, which proves to be the first sign of matter-formation. In our present case, the expansion of gravitation, proceeding from the surface of the singularity new matter is sparkled out as the bubbles in the beer-glass into the surroundings.

In the foam layer covering the core the bubbles of the super-atom are now rubbing against each other, thus producing the first signs of a very low temperature of a few billionths of Kelvin degrees. And this tiny temperature increase causes a Neutron-Condensate to be developed. In the immediate vicinity of the strong gravitational influence of the core, Neutrons are the result of the Neutron Condensat, compared to Protons being the consequence of the Bose-Einstein-Condensate.

During further very low temperature raises, the first neutrons develop from the condensate and begin to oscillate and are driven further and further out by the following foam. The "evaporation" under the core´s surface becomes a continuous process and new Neutron-Condensate is produced, which also creates more neutrons. The course of the evaporation takes place in complete harmony with the opening up of the space by the arising matter. The gradual development of matter takes place shell-shaped around the core, every new stage in the development chain: foam, the following products of creation, to which we are referring later, and the vast Black

Zero Energy, each form a new shell around it. Being the neutrons the first matter-shell around the foam of the neutron condensate.

Neutrons, Protons, and Helium

Gradually more space is available on the way of the matter in to the vacuum. At the end of the half-life period of 11 minutes since the formation of the first neutrons, they begin a continuous decay and split up into 3) Richard Feyman and Harald Fritzsch: *"protons, electrons, and neutrinos"* and the protons soon catch the liberated electrons. This causes that the newly formed protons, with other neutrons present, form deuterium. In the sun and in the stars, the further development of the assembly chain: H to 4He takes place at extremely high temperatures, which are necessary to overcome the rejection of the positively charged nuclei. In doing so, the atoms are brought to a high speed, so they come so close that they now can fuse. This process is not possible in the actual "Gravity-Universe" with its lowest temperatures, near the absolute zero point, and so the problem must be solved with an equally effective and suitable form. In addition to the current hot core fusion, there is another, the "catalytic" or "cold" fusion. In this case, especially at low temperatures, the enormous repulsion forces of the positively charged atom nuclei are neutralized by means of a catalyst in order that they can come so close together that they melt together.

The catalyst is a Myon, which is formed as a subspecies of the neutrino in the case of neutron-condensate decay of the neutron (*Electron- and Myon-Neutrino* = Weinberg 4). Myons are "heavy" electrons; both have ½ spin and negative charge, which is many times greater (200 times) in the myon than in the electron. If a normal atom, whose core is positively charged, captures a myon, which is very happy to do so, it releases the electron and its strong negative charge binds approximately the whole positive charge of the nucleus. Such neutralized nuclei can come so close together that now the attracting core forces become effective. The orbit of the myon around the nucleus is much smaller than that of the electron, and thus a myon atom with neutralized charge, which has no repelling forces, comes so close to another atom that both, at this low temperature can fuse. During the fusing process, the myons are released and they can now initiate further fusion processes. The heavier atoms that are formed now begin to trap again the free electrons. The heavier atomic nuclei are formed from lighter constituents: from deuterium and 1 proton = Tritium, Deuterium and Tritium results in 4He and 3He and a neutron produces 4He.

The two shells of the 3He and 4He are already the precursor of the Intermediate Layer. The prevailing environmental conditions: shortness of space and lowest temperatures, have an advantageous effect on the catalytic core fusion.

The great accumulating and releasing energies are emitted as a radiant energy, especially as gamma quants, into space, G. Hasinger 5): *"In the conversion of hydrogen to helium, very much energy is emitted which is radiated in the form of light"*. And: H.Lesch / J.Müller 6) *"From a certain temperature and density, the protons of the hydrogen fuse to helium nuclei. In this way, energy is released in the form of gamma rays"*. And G. Börner, says 7), *"about 10 billion radiation quanta meet a matter particle in the present state of the cosmos."* Radiation is an integral part of our Universe. It is omnipresent and in combination with the Dark Matter it forms one of the main components of the "Gravity-Universe". For its functions, it needs the respective medium: e.g. For the effect of temperature atoms or molecules are responsible with their vibrations; Light and warmth are only visible or perceptible by the air excitation, so the night is dark and cool, because the air molecules of the earth's atmosphere are not activated by direct irradiation of the sun during the night hours.

The constantly arising rays are now shooting out into the vacuum, opening up more and more space for the following matter. The heat generated by the whole process of the formation of the matter, however, is almost completely absorbed by the enormous cooling temperature of the expansion, so that it can only rise to a very few degrees K during the formation of matter. There is not a more ideal and more perfect cooling system: here the ambient temperature is 0°K and the resulting matter is constantly kept at the lowest level due to the strong heat losses that result from the enormous expansion.

In the present case, no extremely high temperatures occur, which are not necessary here. It must be understood that the temperature can grow only to a small extent, and that it must be able to increase in fine harmony with the conditions of the matter, first because the already existing pressure close to the nucleus does not permit an additional increase of pressure which would arise because of the heating of the atoms. Second, because the prevailing lack of space would not allow this. The actually expected initial product, the hydrogen is initially completely neglected, for the time being it is not in demand in this young Universe. On the other hand, it will play an important role in the future.

In the actually accepted Universe-model, so much hydrogen is formed at this moment that it is condensed in the young Universe and produces the first star generation. These are red giants, which have a very short-live and grow to Supernovae at the end of their lifetimes. These send as soon as they explode a series of heavy elements into space, which are to be used in later solar systems. This step is not permissible in the present model for two reasons: first in the case of half-time neutron decay, so many protons are formed, which, as already described, combine with the neutrons to form Helium 4 and virtually no proton excess remains. This should be so at this time, because hydrogen will only be needed at a much later time when the stars are formed in the galaxies.

And second, the heavier elements that would arise during these first Supernovae explosions would be scattered indiscriminately into the immeasurable vastness of the Universe, and none of the "tiny" stars in the galaxies will attract such dust from such great distances for the formation of their planets, moons, etc. Ultimately, this step means an unwanted and impermissible "pollution" of the Universe. The formation of the heavier elements would only be expected when these emerge in the individual galaxies from the short-lived giant star, f.i. in our Milky Way, whose predecessors have already provided the heavier elements in our solar system. Thus, the remains of the giants are located in "reachable" proximity of the stars to create planetary systems. This last discovery has already been accepted in the recent past, as for G. Börner 8) says: *"Every carbon or oxygen atom on the earth comes from the interior of a star with a great mass, and when the explosion happened, it was hurled into the "interstellar space".*

"Interlayer"

The two opposing components of the "Gravity-Universe", the Primeval Energy Core, and the material realm can't be in direct contact with one another; since they are on one hand the core with the infinite density, and on the other is the space with the infinite emptiness. They can only exist side by side when they are separated by an "Interlayer". The "Gravity-Universe" owes its existence to it. The Interlayer has a completely neutral behavior; it can therefore, be in contact with the Primeval Energy Core and with the matter. These are high demands expected from it. If we consider what can fulfill them, we must admit that there is no other energy that can fulfill this task. There is also no intermediate substance between energy and matter. Therefore, we must seek a candidate with very unusual features in the limited selection among the elements. What shall be the characteristics of this Layer?
- It must react completely neutral on both sides.
- It must ensure the temperature drop from the matter realm to the 0°K of the core.
- It must protect the core from all forces and radiations.
- It must not have a gravitational effect, but it must be completely transparent to gravity.

Looking at these qualities that are expected of a matter, one can't imagine that there is such one that fulfills these requirements. Nature is the most economical, most selective and perfect of all master constructors and has the right answer for every need. Among all elements, there is one which unites these conditions, and this is Helium.

Helium is abundant in nature and is a precious gas, so it has a complete outer shell and therefore does not interfere with other elements. It is a "waste product" from star synthesis and is not flammable. It is neutral to all forces. However, the behavior of helium at low temperatures is important and that is here particularly the case. For this privilege we could think that nature has created this element precisely for the present task. Helium does not have any other notable use in this cold state in nature. Heluim4 has two protons, two neutrons and two electrons, its boiling and evaporation point is $4.2°K$; at $2.17°K$ is the freezing point, He4 is superfluous and remains so until $0°K$ without freezing (it becomes solid only under pressure of more than 29 bar). This superfluid behavior is called the fifth physical state. The properties of the superfluid Helium are extraordinary: it completely loses its viscosity; it flows without energy expenditure and without frictional resistance. At low pressure, helium loses its gravitational force. He4 has an isotope, Helium 3.

The behavior of the two atoms in the superfluid state is different: He3 becomes liquid at $3.2°K$, but only at $2.4°K$ it is superfluous, and also loses its gravitational force. It can be found sparsely in nature. He 3 cools down better and becomes solid at 34 Bar. Now we have to imagine that the Interlayer, which separates the outer space from the core, consists of He4. On the inside, it is provided with a thin coat of He3 facing the core. The reason for this assumption is that this thin coating significantly improves the superficial properties of both components and He3 is particularly well suited for its specific task, and it is even better than He4, as contact medium to the core. He4 is a whole-spin Bosom whereas He3 is a half-spin Fermion and isotropic and requires higher pressure to become solid. Both layers improve the expected barrier and insulating properties of the Intermediate Layer.

He3 has an extremely low phase transition; it becomes superfluous at $2.4°K$. This is interesting, since this thin He

3 coating is the first real matter-shell on top of the super-atom- foam and the Neutrons which cover the core.

Now there is another factor which substantially improves the task of the Interlayer, namely the so-called "sticking". As already mentioned in the cold fusion, a myon occurs as a catalyst which binds the He atoms together. This myon-electron has a very short live, it has a lifetime of 2 billionths of a second. In that short time, a myon can produce up to 2,000 merges if there were not this sticking. Sometimes the myon is not released after the fusion of the He atom. This happens less than one percent of the cases (perhaps far less time under these extreme conditions). And it is precisely this more compact and heavier helium atom that is neglected by the gamma radiation, which carries the normal He atoms into the Universe. And preferably these heavier He-atoms form the Interlayer. This gives it a well-established stability, which improves its function as an isolating component between the core and the outer world.

And Helium, Helium, Helium, the wonderful pure, and so much desired precious gas,(which now hardly reaches two degrees K), flows - it is super-fluid and weightless – carried by the radiation and driven by the emptiness of the vacuum, unrestrained into space. Helium 4 is superfluid from 0°K to 2.17°K, and in this state it has a very special peculiarity at low pressure, which prevails at a certain distance from the nucleus: it is weightless! Its isotope 3He behaves in the same way. The great gravitation from the nucleus does not influence Helium gravitationally. The idea suggests that this element has been specially created for its task as an Interlayer. And in its initial conquest of space, its weightlessness is an incomparable advantage. It might be said that it is driven out by the just emerged radiation in order to conquer the Black Zero Energy as Dark Matter. This weightlessness was already advantageous to it at the very beginning of his long voyage: since the initial lack of space near the nucleus, the individual He atoms were very close together and did not interfere among each other and they separated easily when they started the trip into the space. The uniform distribution in the widths of the space also happens without any problem.

THERMOMETER

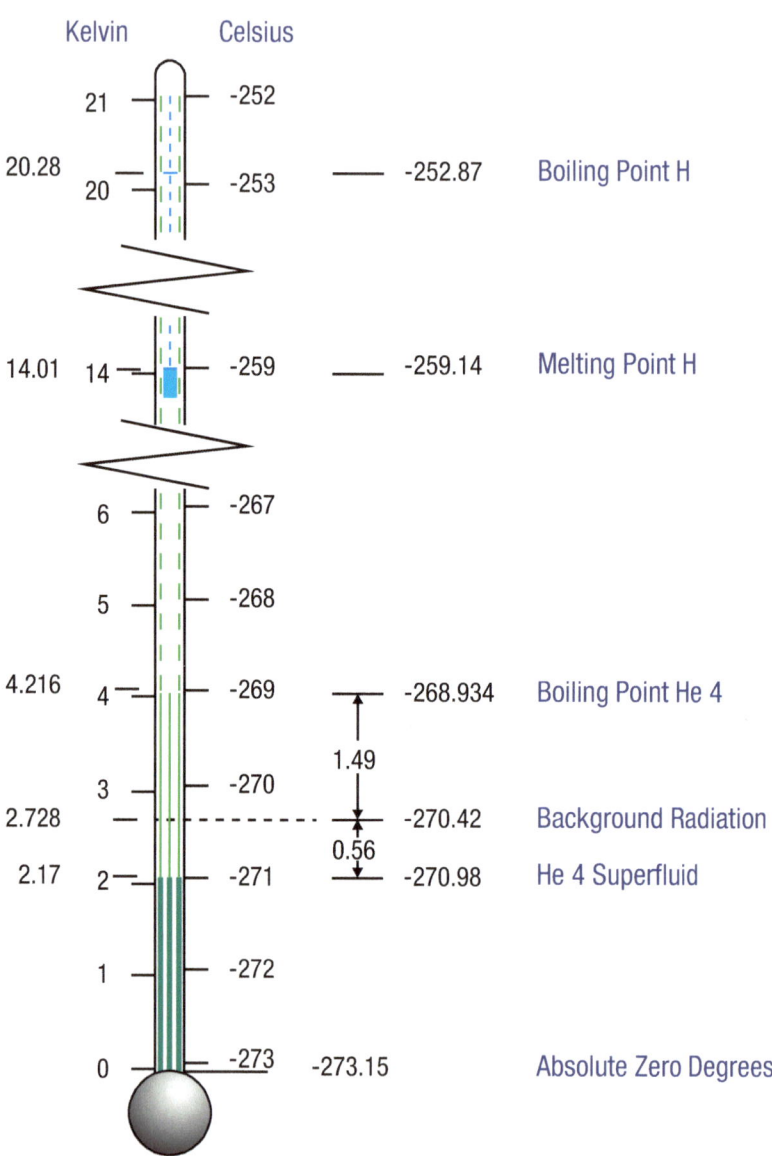

Drawing 3

Once the atoms are sufficiently far away from the core, and the prevailing pressure is diminished, the supra-fluid helium regains its gravitational power, and behaves like any other matter.

During the further development, which the Universe undergoes in its formation, Helium will give up its supra-fluid state, which is now no longer in demand in the vast Universe and gamma-rays of the surrounding radiation heats it to 2.73°K. This temperature is now always maintained, and Helium prevails from now on in liquid form.

When the energetic gamma radiation began to exist, their descendants - (The more space was available, the larger the length of the waves) - the entire young radiation, had a very long journey to go. *"Radiation does not need for its spreading an intermediate carrier"* - G.Börner 9). The farther away the radiation from the nucleus was, the more space was available and so their frequencies decreased and the waves became longer and longer. *"Certainly it maintained its speed of light, but its energy gradually decreased"* P.Körning 10). Nevertheless, it drifted farther and farther into the vacuum and was followed by the subsequent helium current, and so both conquered further new territory. After many billions of years, the energy of the radiation became smaller and smaller, and the distances between the crests of the waves increased, and finally it was held by the outermost edge of the gravitational field which originates from the core, ending finally in giant circular orbits on the horizon of the Universe.

Thus, the first and greatest "hunger" of the vacuum-giant is not nearly satisfied, but diminished until the corresponding necessary density has been established.

Apparently, an exact quantification of the helium in the intergalactic space has not yet been established, for S. Weinberg 11) says: *"Of course, we are still not really sure today that the frequency of helium in the Universe is in the range of 20-30%".* And H. Genz 12) says *"How much matter there is in space between the galaxies is unknown".*

In any case, the so sought -for and missing Dark Matter in the Universe is covered by liquid helium and this probably accounts for the missing 80/90% of matter.

The greed of the vacuum, the emptiness of the Black Zero Energy, which is constantly acting upon the stream of matter which is emitted from the nucleus, will never be saturated, and it represents a requisite for the function of the Universe.

After billions of years, the tension between the gravitation of the nucleus and the emptiness of the vacuum begins to be reduced by helium, and this is reflected in the helium production, so that it decreases slowly to the extent necessary, in accordance with a controlled and well-coordinated sequence. With the throttling of the helium production, the distances between the individual protection shells near to the nucleus diminish, and they move closer together, towards the nucleus, and into a more concentrated gravitational region. And here the newly emerging helium atoms are retained in the "vicinity" of the core, forming a denser shell, the Interlayer. The Interlayer belongs energetically to the control area of the energy core, and the actual space of action of the vacuum begins only outside of it.

This Interlayer can then protect the nucleus from life-threatening expansion. This marks the end of the first step of the formation of the Universe. In the birth hour of the Universe, we know, time and space were created. But another very important component is added: the temperature. It plays an equally important role in the Universe.

The next step that now follows is the population of the Universe with galaxies and ultimately the take up of the final functioning of the same.

Galaxies conquer the Universe

The "Gravity - Universe" completed now its basic structure, there is no more obstacle for the further completion of the same. In order that the now missing galaxies, stars, planets, and any matter present in space are formed there exists a need of further transformation of energy into matter. This takes place in such a way that vortexes are formed in the helium Interlayer. Supra-fluid helium has the property of forming eddies easily. Such a vortex abruptly cuts out a "Part" of the surface of the core, and this "Part" is now thrown into space by the vortex with unimaginable force. This whole process is also strongly supported by the Black Zero Energy– the emptiness of the vacuum. The vertebra ebbs immediately, and the spot on the core heals at once. And far, outside of the Interlayer and the strong gravitational zone of the nucleus, this "Part" which is still concentrated core-energy, travels without limit of speed (liquid 4He puts no resistance and energy is not bound to a speed limit). The "Part" is now heated by the surrounding temperature of 2.73°K prevailing in the outer space. At a certain distance of the core it starts to expand and slows down to move just under the speed of light and since it is a "naked" singularity, it begins to "melt". This means that matter is now generated from primeval energy, in a violent however controlled action of the phase transition from one to the other Physical States. Practically here we see a replica of the "Big Bang".

This expansion process usually produces a strong radiation and luminosity, over the entire width of the spectrum, above all gamma bursts and emissions of radio waves. And with the continuation of expansion, the high initial speed further decelerates, which now drops to 240,000 km per second. The structure is called a "Quasar" whose luminosity surpasses other known radiation-objects in space, and from which a new galaxy is born, and this is how it happens: The singularity -"Part" has met completely different environmental influences than those which existed during the evaporation process of the Primeval Energy Core when it became active. Here, the energy "Part" finds a far lower gravitational pressure and here exits a temperature of 2.73°K. And these external influences are causing the "Part" to start "melting".

During the expansion a Bose-Einstein-Condensate is produced. And just a little bit of Neutron-Condensate emerges. Both condensates form at the lowest temperature, (fast-running phase transitions have a strong expansion cooling) Super-Atoms, which produce both protons and some neutrons. Due to the low prevailing pressure from the outer world, now mainly protons and very few neutrons are produced.

The strong emerging radiation at the same time warms the "newborn" atoms, but the strong expansion constantly cools the resulting matter. Quasars thus have an expansion cooling system which acts exactly as already described in the process of helium formation near the nucleus. The main product, as we have said, is hydrogen; This is confirmed by P.J.E.Peebles 13): with respect to the quasar PKS 0458-02: *"Its optical spectrum has a prominent feature identified as the Lyman Alpha resonance line of atomic hydrogen ... in a gas cloud along the line of sight between us and the quasar"*. It is to be assumed that the hydrogen in the initial state has a temperature above 15 ° K, and is therefore still gaseous, and together with a small amount of helium produces a gas cloud which, as a result of the self-gravitation, gradually condenses and little by little adopts the ambient temperature of about 2.73°K.

At this temperature, the hydrogen is in solid state, forming a dust. The liquid helium, which is produced during the process, damps some hydrogen dust in the now formed gas and dust cloud, it begins to "lump". The lump begins to rotate, and like in an avalanche, glues more and more hydrogen dust from the cloud to the lump, attracted by its gravity, so that it becomes bigger. The greater it grows, the more pressure builds up in its inside until this becomes so great that finally a nucleus synthesis is triggered in the interior of the lump, in which hydrogen is converted to helium - a star is born. In this way the hydrogen cloud brings about more and more stars from further clumps and stars and the cloud create a new galaxy. Meanwhile, the speed of the cloud has drastically reduced to 100,000 km/sec.

Now begins the colonization process of the Universe, and the vacuum is enriched by quasars and by their finished product, with galaxies. The expelled "Parts" of primeval energy, from which quasars and galaxies are formed, have more or less the same amount of energy, so the galaxies also have a quite uniform size. This population of the Universe continues for billions of years, until the Universe is provided with a sufficient galaxy population, the final objective is reached and theoretically the once captured second singularity has now been reduced and transformed into matter. The equilibrium state between the original single singularity in the nucleus and the now present matter in the Universe is reached the nucleus and the matter in the Universe hold a gravitationally equilibrium. However, the emptiness of the vacuum will never be satisfied, it provides the so called "Black Zero Energy" which is a special support for the migration of the galaxies from the nucleus into the empty space.

The Energy Circuit

The equilibrium state of the Universe, which has now reached the nucleus and the vacuum, through the increased generation of matter, will not stop the further formation of new galaxies. However, in order to preserve the assumed balance between the core and the external matter forever, matter in form of energy from the matter realm must be recalled. The cycle: the formation and disintegration of matter, which is now a continuous flow in the Universe, will never stop.

How many billions of years the entire population of the Universe lasted can't be answered? Today's assumption of the age of the Universe of 14 billion years is completely underestimated. If the average age of stars in the magnitude of our sun amounts to 9 billion years, then countless generations of stars must have existed throughout the entire stellar occupation of the Universe. We see, therefore, that the "Gravity-Universe" must have an unimaginable broad and spatial extent, and that its age is far, far higher than assumed today. A galaxy is, in the initial stage, a large, bulky dust and gas cloud consisting of hydrogen powder and, to a lesser extent helium damp. The hydrogen cloud condenses over time, driven by self-gravitation and constant cooling, and begins to rotate. Hydrogen solidifies to stars. As a result of the prevailing gravitation and rotation, the galaxy now becomes a disk shape and on both sides it is still surrounded by halos of hydrogen and helium (dark matter). The last stage is the spiral galaxy, which spreads its arms far into the intergalactic space.

In the course of time, a black hole is formed in the center of the large spiral, into which the eddy current flows, and into which everything is absorbed which is near this vortex: matter disappears: suns and planets, radiation and whatsoever in the perimeter of this hole. This process is a compression process, in which matter that is carried along is becoming increasingly accelerated and condensed. The resulting increasing pressure atomizes everything that enters into this flow.

The constantly smaller particles become ever heavier and are further compressed until even the electrons are forced into the atoms. Also the radiation loses its spread energy due to an increasing lack of space and finally flashes as gamma radiation before it disappears in the hole. Before the inevitable end, a drop in temperature takes place, since the atoms lose increasingly vibration space. This cold zone, together with the still increasing pressure, favors the ultimate compression of the matter. *"The colder gases and dust get, the more easily they are compressed"* (M. Begelmann / M.Rees) 14). The temperature of 0°K is reached during the phase transition.

Finally, precisely at the edge of the event horizon, the compressed matter stream will reach the speed of light and there the matter will be transformed subsequently from the neutron concentration and the following super-atom into an enormously increased gravitation - a singularity. This corresponds to the process already mentioned at the beginning about the physical state - transition of the neutron into the singularity. This Primeval Energy, which is continuously generated in the black hole, is absorbed by the singularity, the Primeval Energy Core, in the center of the Universe instantly and at an infinitely great speed, H. Genz, 15) says*: "If there is no "intermediate – carrier" of gravity, its effect must spread infinitely fast."*

THE SLIDE

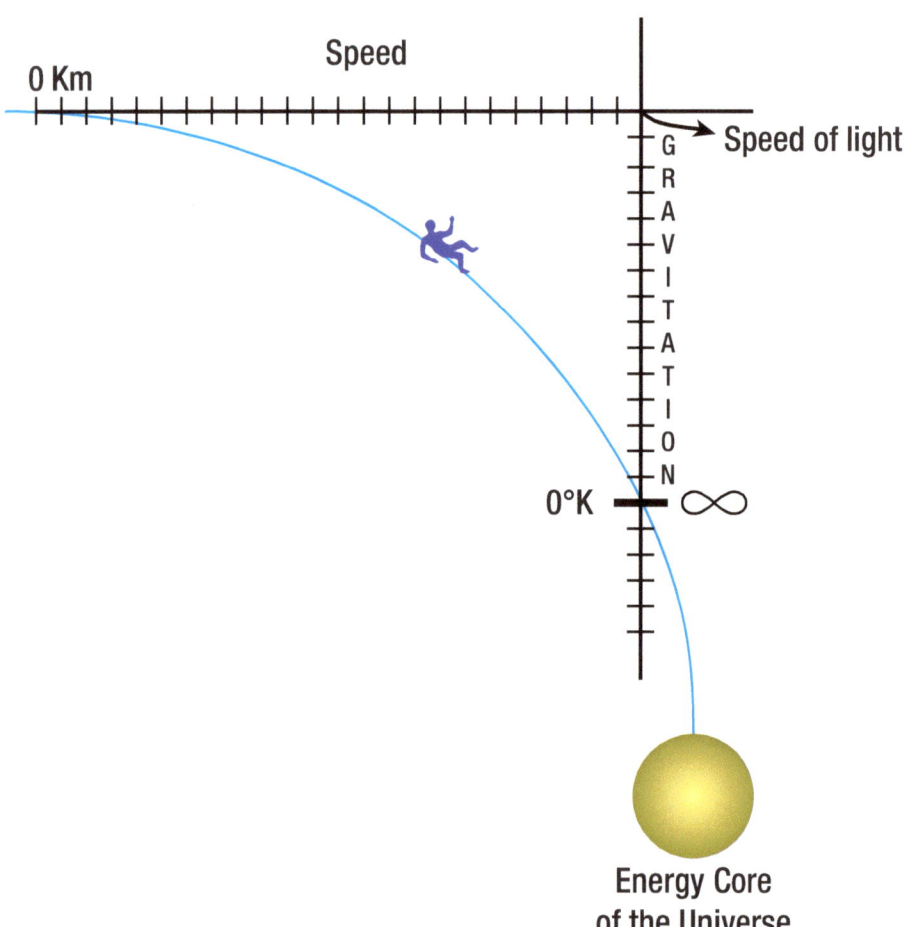

Speed

0 Km

Speed of light

GRAVITATION

0°K ∞

Energy Core
of the Universe

Drawing 4

... And for the sake of importance, we shall recall this process in detail and in a pictorial manner. Just as everyone likes to flit a flat stone over the small combs of a lake, it might also be tempting to throw a stone into a black hole, and that is exactly what we want to do now. For this we will roll up the black hole and stick it to the wall, on the upper left is the big opening and below on the right is the small hole. And now we want to paint a coordinate system on our hole: the speed arrow goes horizontally from left to right and shows the respective speed. Where the arrowhead is, the speed of light is reached and exactly there we draw the vertical gravitational arrow with the ever-increasing gravity down, since everything falls from the top downwards into the hole. Right at the arrow point, the gravity is infinite. Now we draw a curve from the beginning of the horizontal arrow down to arrow head of the vertical arrow. Let us assume that this curve is a slide. Now we *put a* stone on this slide and want to prove what Mr. Einstein has predicted - namely that all objects become heavier with increasing speed. So it happens: the stone goes and starts to move, then to slide, chase, rush, fly, shoot ... and it becomes thereby heavier and heavier, and it is more and more compressed ... the ride goes ruthlessly, the speed is getting faster and faster , the stone is pressed further together, the weight increases rapidly, but the ride gets even faster, the stone is compressed further, the stone is rushing down the track, has greater and greater gravity - we now replace the little word "has" and say it "becomes" increasingly heavier, - and the stone is only a point, but it weighs as much as a whole skyscraper, - but now it accelerates the last meters down the way, to the point of intersection of the slide with the arrowhead and precisely at this point the speed of light is reached, it is just crossing the end of the track ... yes, and there the inevitable happened... our stone is no more, it has at the transition to the light speed undergone a phase transition from one to the other physical state, and what once was a stone, or matter, has now become pure gravitation. But the journey is not finished, our gravitational-stone, that is, the gravitational "Part", falls, like the apple of the tree, directly into the nucleus of the Universe, crosses the intergalactic space smoothly, one of the properties of the supra-fluid helium and has instantly merged with the nucleus. Thus the stone was transformed into the most pure gravitation!

And it gets not only in the free fall into the hole, but it is just jetted into it. Just below the head of the vertical arrow lies the 0°K point. The stone was not only becoming heavier, it was getting colder on its descent. The big, honorable man was right, but he stepped out too early! This is understandable, since 100 years ago he could not have foreseen that something exists after the speed of light - only matter is bound to the limit of speed of light, - energy is not.

However, we can repeat this nice game with all sorts of things: a house, a star, a planet, a neighbor's dog ... but let's put a handful of photons on the slide, at the beginning they have normal wavelengths but they become shorter and shorter waves as the acceleration increases and to finally radiate as gamma radiation, just before the end of the track they change over at the point of intersection to Gravitational Energy, just like our stone.

In summary, we can state:

Every object that accelerates at speed of light also will become infinitely heavy and in order to achieve that an infinitely great force is therefore required. How does the black hole solve this problem? As the acceleration increases, the matter is more and more compressed and becomes increasingly heavier. The phase transition from matter to gravitation starts before reaching the speed of light. The matter gets smaller and smaller, so it loses mass and transforms it increasingly into energy or gravitation. This accelerates easily to the speed of light. In addition, gravitation has no temperature, the more gravitation is produced, the cooler the matter. Thus, before reaching the speed of light, it changes into gravitation and completes the transition when reaching the speed of light. Thus Matter was transformed in Energy. Now we are at the end of our journey, now the big cycle is closed. - Thanks for your company! Wasn´t it an interesting trip?

More Reflections

To the world-wide validation of the force of attraction, which is applicable in the matter ambient T. Berry and B. Swime 16) will enjoy us with their thoughts: *"Newton saw gravitation as the primary force, that holt together the so vastly expanded Universe. And most important he showed that the laws of gravitation, which we know here on earth, also apply to the entire extraterrestrial physical world, including the celestial objects observed by our astronomers. Once it was proved that such a universal cause of order acted in the Universe, the researchers who devoted themselves to other natural sciences could assume that they are already on the right way."*

In the "Gravity-Universe", we experience a very focused and well-balanced process that does not tolerate extreme excesses. Nature is the most economical and most meticulous and conscientious administrator achieving in all the processes what is required in every case, taking into account and utilizing the smallest possible energy expenditure. Massless debauchery such as unnecessary high temperatures, the destruction of matter by antimatter, the abrupt stopping of a normal process, and other anomalies or wastes, do not belong to it. This fact must be kept in mind during every observation of the "Gravitational Universe". Besides of helium, also other particles, such as neutrinos, electrons, photons, etc. form part of the dark matter in space. Helium is constantly held at 2.73°K by the surrounding radiation field in the intergalactic space, so that it remains continuously in liquid state, and every smallest excess of heat is returned immediately as the observed radio-wave-emission. This is exactly what we see in the case of hydrogen, which also emits radio-waves, namely on the 21 cm wave, and according to P. Padmanabhan, 17) this happens to be so: *"The n = 1 state of the hydrogen atom, in which the electron and the proton have parallel spins, has a slightly higher energy than the state with antiparallel spins. The radiation emitted when the atom makes the transition between the two states, has a wavelength of 21 cm ..."*

Such experiments have already been carried out in the Physical Department at Harvard University. It has already been shown experimentally that 3He also "oscillates". Apparently no such attempt has been made with liquid 4Helium. It is no doubt that it also "rushes".

This means that 4Helium also undergoes a continuous change: warming, by means of environmental radiation, falling back into a reduced colder level with simultaneous emission of the excess energy, - warming up - cooling with energy release, etc. -- In this way 4Helium keeps the Background Radiation constantly at such a uniform temperature.

As a precious gas, it does not couple with other elements and it has a very important task. It acts as the great isolator in space, which keeps the chemically active hydrogen under control. The fusion processes of the stars and their eruptions, quasar processing and supernova explosions in the Universe also proceed safely because of the presence of helium, so that they can't trigger a catastrophe.

Helium is constantly regenerated in the stars since helium is also devoured in the black holes during the decline of matter.

For mankind the solar radiation has always been of vital importance. But, here the question: is the solar radiation the main product or just a by-product of the fusion process in the sun? The neglected helium has probably the real privilege. This helium continually renews the stock in the galaxy, since the helium is also swallowed in black holes. Large part of this element also goes into free space, where it has to keep up with the stock, which is also affected by black holes. Where helium is produced, there is also gamma radiation, which carries the helium vapor far into space, where it cools down to 2.73 degrees K and incorporates the dark matter. Here it keeps this temperature for ever and is permanently liquid.

Constant observations of the Universe have shown that galaxies are structured into larger groups. An explanation for this would be that matter, the farther away it is from the gravitation of the nucleus core, have to move together into tighter formations in order to compensate the gravitational diminution, which arises when the distance between the nucleus and the matter increases. On one hand, there are conurbations of galaxies, on the other hand, voids, where the matter is very scarce. Voids are "low-pressure areas", which in turn stimulate the formation of vortexes on the Intermediate Layer, so that there emerge more quasars and an increased formation of matter occurs. So the farer we come out into space, the more reduced galaxies merge together and there at the utmost edge of the space the hungry black holes have devoured al matter, radiation, but also all remaining helium is transformed in Primeval Energy which is absorbed straightaway by the nuclear core. Here is the limit of the "Gravity-Universe".

In order to understand our Universe correctly, we must not look at it from the matter side, as we do, but from the actual center of gravity, the energy side or the singularity. Although, our physical rules and laws are not applicable, but everything in the Universe was once concentrated and hidden there. The vacuum did not have anything to offer, it became only with the matter which filled it, part of our world. If we lose from the fetters of the classical thoughts, we understand the whole Universe correctly, and then we also understand that, in our material realm of the world or in the singularity of the core, everything is nothing but gravitation. The coordination of components, already point to a common denominator.

Thus, the three forces bound to matter, namely, nuclear power, radioactivity, and electromagnetism, are nothing else than part of the gravitation in one of its physical states.

Conclusion

Finally, K. Ferguson 18) gives us one of her beautiful thoughts: *"From Hipparch we know that he always preferred the most uncomplicated hypothesis which reconciled with the observations, and Neoplatonism favored the scientists of the Late Medieval to trace in nature for simple mathematical and geometrical regularities. Copernicus was firmly convinced of the higher harmony of this order and saw in it a strong argument in its favor. Kepler's work was also inspired by the search for such harmony. Some scholars have expressed this preference in a religious way: God's creation: NATURE is a work of great elegance and simplicity, and we must try to understand and explain in the same way."*

BIBLIOGRAPHY

1) Gerhard Börner — Schöpfung ohne Schöpfer? P. 18
Creation without Creator?

2) Web — www.factbites.com/topics/Neutronium

3) Richard Feynman and Harald Fritzsch — Sechs Physikalische Fingerübungen P. 94
Vom Urknall zum Zerfall P. 201
**Six physical Finger- Excercises
from the Big Bang to the Collapse**

4) Steven Weinberg — Die ersten drei Minuten P. 137
The first three Minutes

5) Günther Hasinger — Das Schicksal des Universums P. 84
The Destiny of the Universe

6) Harald Lesch/Jörn Müller — Kosmologie für Fussgänger P. 183
Cosmology for Pedestrians

7) Gerhard Börner — Schöpfung ohne Schöpfer ? P. 74
Creation without Creator?

8) Gerhard Börner — Schöpfung ohne Schöpfer ? P. 177
Creation without Creator?

9) Hoimar v. Ditfurth — Im Anfang war der Wasserstoff P. 85
At the Beginning was the Hydrogen

10) Gerhard Börner — Schöpfung ohne Schöpfer ? P. 14
Creation without Creator?

11) Peter Körning — Auch Genies können irren... P. 154
Also Genuses can be wrong

12) Steven Weinberg — Die ersten drei Minuten P. 177
The first three Minutes

13) Henning Genz — Die Entdeckung des Nichts P. 375
The Discovery of Nothing

14) P.J.E. Peebles — Principles of Physical Cosmology P. 92

15) Mitchell Begelman, Martin Rees — Schwarze Löcher im Kosmos P. 70
Black Holes in the Cosmos

16) Henning Genz — Die Entdeckung des Nichts P. 199
The Discovery of Nothing

17) T. Berry und B. Swime — Die Autobiographie des Universums P. 240
The Autobiography of the Universe

18) Padmanabhan   Structure Formation in the Universe P. 70
Structural Formation in the Universe

19) Kitty Ferguson   Das Mass der Unendlichkeit P. 119
The Measure of the Infinit

EPILOGUE

The present Universe model did not emerge from a momentary impulse, but is rather the product of years of studies and profound reflections on the matter, in order to find an accurate answer to the many open questions on this subject. Besides, important scientific knowledge has already been taken into account when formulating this composite. It is fundamentally based on the precise understanding of gravitation. With this clear definition of gravitation, this model could be understood and explained successively.

As early as 2002 the first attempt was made to tackle the issue. From today's point of view, however, it is insignificant because at that time the so important interpretation of the gravitation was at that moment not clear. In 2004, when the present determination and the extensive significance of gravitation throughout the Universe were detected, the next step could be approached: The Structure of the Universe. This essay in which the nature of the Universe was formulated, describes not only its integration, but also its function.

When this task was finished the justified question came up, if it ever could happen to understand and describe the most important section of the origin of the Universe. At that time there was not the slightest belief that this topic could ever be met by any person. However, in the following years thoughts matured and gave an ever more complete initiative until the rough draft was finished in 2012. The individual details emerged more and more clearly the further the formulation progressed and the intensive study of the subject completed the picture. Thus, the present work, the "Gravity-Universe", emerged - one thing leading to the other. The previous study, which concerns the construction of the Universe, is superfluous, since in the present and most important component of the work, the structure and the functioning of the Universe is sufficiently described and hopefully well understood.

This Universe model explains so many open questions "out of itself", and it could be said that the answers are appropriate and they fit right into the master plan of the environmental picture. This privilege gives it a great reliability. Still each new Universe model can only be taken serious if it involves gravity in some form. All models which are known actually; may not be viable in the long term.

The End

Content

Autobiography

Naval Officer Franz Sundfeld, born 1924 in Mexico:

1931	German School, Guatemala.
1939	Physic-Teacher explains the Big Bang. My interest in Cosmology was stimulated.
1943	Naval Academy Flensburg: Astronomic Navigation. Special training: sun, stars and ships position.
1998-2019	Self-study Cosmology, reading many scientific books. However, all were based on the same dead cosmos model.
2002	First attempt to explain the living Universe.
2004	Identification of Gravitation as a universal force.
2004-2019	Formulation of the "Gravity-Universe" concept, which is a self-renovation system and full of live.